INTRODUCTION
TO
INTERNET PROTOCOLS

THEIR ARCHITECTURE, THEIR PROTOCOLS
AND THEIR FEATURES

PHILIP AVERY JOHNSON

THE COLLEGE OF WILLIAM AND MARY
AND
FLORIDA INSTITUTE OF TECHNOLOGY

iUniverse, Inc.
New York   Bloomington

iUniverse books may be ordered through booksellers or by contacting:

iUniverse
1663 Liberty Drive
Bloomington, IN 47403
www.iuniverse.com
1-800-Authors (1-800-288-4677)

Because of the dynamic nature of the Internet, any Web addresses or links contained in this book may have changed since publication and may no longer be valid. The views expressed in this work are solely those of the author and do not necessarily reflect the views of the publisher, and the publisher hereby disclaims any responsibility for them.

ISBN: 978-1-4502-1675-3 (sc)
ISBN: 978-1-4502-1676-0 (ebook)

Printed in the United States of America

iUniverse rev. date: 03/08/2010

To my wife, Margaret, and my children, Lori and Jeffrey.

Foreword

I OWE A LARGE debt to a number of people over the years. I particular, to Mark Klerer, my boss and mentor at Bell Laboratories, who introduced me to the topic of interworking. Mark is the most unique individual I met at Bell Labs: his history says that he is a survivor in whatever environment; he was born during 1945 in Berlin as a Jew. I don't know how he survived, but he did, became a displaced person (DP) in Europe, made it to Great Britain, was educated there, came to the United States and started his career at Bell Labs where I met him. I can still see him working on his computer, doing compiler research, handling a phone call, talking with a number of people, and doing other things, all at the same time. I've forgotten when he left Bell Labs; I think he went to Nortel for a while and then I lost track of him. But, what an inspiration in many ways. He was not afraid of rocking boats. As a boss, he told me that he would defend my decision on any technical matter, just let him know, wherever in the world he happened to be at the time. Working for him was a pleasure.

Others that should be acknowledged include: Ben Avi-Itzhak of

Rutgers, Larry Pulley of William and Mary, Bob Beebe of Verilink, Chuck Huffman of Rockwell, Catherine Elder of Florida Tech, and Bill Faust of Bell Labs. And many who I've rubbed shoulders with over the years.

Contents

Chapter 1

Introduction

WE EMBARK ON THIS brief book by indicating the basic principles that we will be using to develop the internet protocols and their properties. Three main principles are formative in this effort: 1) the client – server concept, 2) the layering concept, and 3) the concept of hiding details of underlying hardware details while providing universal communication services.

The first of these concepts, the client-server, is now discussed with some background, and then discussing message-passing in detail. It should not be surprising that the original idea of computer design is incompatible with communication and networks. Originally, the computer was a single machine, derived from the von Neuman "finite state machine". This machine was a piece of magnetic tape, being fed into a tape reader/writer with a control box attached. This control permitted the one character, currently at a position to be read by the controller, to be read, and would then instruct the tape to be moved forward, back, or remain in position and would then introduce the next control instruction, for the next possible movement of the machine. This model of a computer has remained intact, primarily by theorists. It wasn't long, in practical terms that the idea of getting data to and from the computer was resolved. The keyboard, mouse, printer and so on led to the integrated processor design that we now think of as a "computer". The concept of an operating system gave rise to the notion that the computer was really a network of various devices, controlled by operating system interactions with the devices.

Actually, the framework of processor-operating system-peripheral devices has undergone many technological changes over the years, evolving to the computers of today. However, with the advent of the internet and the web, we have the new networking needs of today, leading to client-server.

Chapter 2

Client – Server

CLIENT-SERVER (ALSO KNOWN AS Manager-Agent), or the client-server paradigm, has seen great growth since the early 1990's. Much of this growth has been supported by a number of industry segments: 1) the application- and data base providers, 2) the operating system vendors, and 3) the workstation vendors. As the client-server offers system expandability without major retrofit and major expense, the database and applications vendors signed on quickly. The operating system vendors were able to reuse tactical design for software solutions: for example, the print spooler in many modern operating systems, which closely resembles the Remote Call Procedure.

2.1 Definition of client-Server

The client-server paradigm connects at least two computers across a network through a protocol of sending messages back and forth, which define what kind of communication will be allowed at this instant. The "client" computer is the processor is the source of messages for aid in processing the user's request. The "server" computer is the processor, which responds through messaging to the "client" in a predetermined way, either by doing the complete job requested by the client or by doing a portion of the processing or something in-between. The amount of work downloaded from the client to the server is determined at the initiation of the communication and can be categorized into the following:

1. Client-based processing (also known a "fat client")
2. Server-based processing (also known as "fat server")
3. Cooperative processing (also known as "pier-to-pier")

Fat client represents processing done by the client machine; fat server represents processing done by the server; pier-to-pier is a split of processing between client and server machines, done for optimal performance reasons.

2.2   Characteristics of client-server

Reasons for the development of this paradigm include: the flexibility of this computing approach has allowed many industries to gradually grow from one computing environment into another, in a graceful (i.e., a non-business-impacting) way. Thus, the end-user can remain quite comfortable using applications on the client-server environment, where changes can be imbedded on the server. This kind of flexibility has also given end-users great flexibility in upgrading to a faster environment: the user negotiations remain on the client with upgrades going to the server. This has also led to the idea of very small/inexpensive machines, called "appliances", which has limited functionality, allowing internet communications. This compartmentalization of the computing resources is typically referred to as modularization (also known as "openness").

Many businesses tend to think of these concepts as a way to emphasize corporate databases, network management and other utilities, normally on the server. The implicit networking with the client means that support for these various functions can be centralized into one (or limited) places; this aids maintenance and ongoing operations of the system as a whole.

2.3   The concept

The model of client-server came from the central ideas in operating systems development: the concept of message passing between client and server; communicating who does what is directly analogous to the development of print spooling in general operating systems.

What is "spooling"? This concept allows a special software tool to initiate communication between operating system and the printer controller (generally, a part of the "printer") to cause the printer hardware to print a user-specified document. The key aspect is that the operation is transparent to the user – actually, the operation is activated by user-specific code, which enables the special operating system code to activate the printer hardware. A more detailed printer spooler operation now follows:

1. an application program performs an "I/O" instruction

2. the I/O instruction activates an operating system primitive command called a "trap"; This is the only way that an application program can communicate directly with the operating system, other than a "call"

3. the "trap" collects various parameters about the "I/O" and puts them into a stack. Examples of the parameters are: 1)filename, 2)number of copies, 3)margin details, 4) font, and 5) color details. At times, these parameters are added to by the user through the user interface.

4. When the operating system scheduler allows it, the stack is sent to the printer controller

5. When the printer controller detects that the printer is free, the stack is sent to the printer for execution

6. At the end of execution, a message (usually, a

parameter) is returned to the operating system.
This message indicates either success or a code for
failure, if the execution was unsuccessful.

Two aspects of this detail are noteworthy: 1) the execution of the
printer is completely independent of the end-user, and 2) the actual
mechanism of the communication between the operating system and
the printer controller has been glossed over in the above. The idea of
end-user transparency and the mechanism for communication form
the basis for message passing.

2.4 Message Passing

In this section, we develop a process for message passing.

— THE PROCESS

We arrive at a place in an end-user applications program where we
wish to do a piece of work. We will generally execute a "call" to transfer
control to the operating system. As far as the end-user is concerned,
program execution is still continuing in the currently executing
application program.

The interrupt handler of the operating system interprets the "call"
instruction: information is gathered by the operating system to keep
track on where to continue processing. This information comprises
data, pointers to specific code and other parameters in the application
code; this information is placed in a stack in the operating system. The
process now follows:

1. transfer of control within the operating system to the kernel
    (of the operating system) occurs. The sole function of the
    kernel is communication. The kernel determines the location
    (specifically, what machine is to perform the work), creates
    a message which encapsulates the stack, and sends the
    message containing the stack to the destination machine

2. Once the destination machine receives the stack in its kernel, the stack is stripped of its message envelope, is forwarded to the interrupt handler, and is scheduled for execution. Upon execution, the results of the execution are placed in a stack in the destination machine's operating system and are returned to the kernel.

3. In the kernel, the results are placed in a message and are returned to the initiating machine. This message is received in the kernel of the requesting machine.

4. The kernel of the requesting machine then removes the stack and transfers the stack to the interrupt handler, which then schedules the application program to return to the place at which it gave up control. The results are then returned to the application program for further direction of processing by the application program.

2.5   Remote Procedure Call (RPC)

The above procedure of using the operating system facilities to bundle necessary parameters, to construct messages for another system, and to transmit the message to another system, whether client or server, has the end-user transparency and communications properties discussed earlier. The above procedure is directly analogous to the Remote Procedure Call of operating systems, as both use the same mechanism.

In terms of actual communication, a couple of things, earlier glossed over, should be clarified. First, when the client is ready to send the message to the server, how does the system know the name/location of the path in the network to use? The client-server paradigm uses a transparent mechanism (to the end-user) to specify the destination address. Although there are three mechanisms that conceivably do this, only one is practical. This is called the Domain Name Server (DNS). The DNS is an additional server appended to the communications network of the client. This server acts as a directory reference for the client, by retrieving machine addresses of servers, able to do the client's work. It works as follows:

1. The client sends a message to the DNS to find a server, which has the capability of doing the client's work

2. The DNS sends back a reply to the client with the machine address (the physical address) of a server that can perform the work

3. The kernel of the client appends the machine address to the message directed to the server for processing.

The second issue for the client to decide whether to adopt a blocking or non-blocking primitive in its operating system. The difference is whether to buffer the message as it is being sent from the client that processing can continue at the client or not buffer and hold processing until acknowledgement of message receipt at the server is received. In almost all cases, the non-blocking option is preferred.

Chapter 3

Layering of Protocols

3.1   Introduction

The purpose of this chapter is to consider the structure of the software located in hosts and routers that carries out network communication. It discusses the concept of layering, shows how layering makes IP software easier to understand and build, and traces the path of datagrams through the protocol software they encounter when traversing a TCP/ IP internet.

3.2   Why multiple protocols

Protocols allow one to specify or understand communication without knowing the details of a particular vendor's network hardware. They are to computer communications what programming languages are to computation. Like assembly language, some protocols describe communication across a physical network. For example, the details of frame format, network access policy, and frame error handling describes X.25 communication. Similarly, the details of IP addresses, the datagram format, and the concept of unreliable, connectionless delivery comprise the Internet Protocol.

Complex data communication systems do not use a single protocol to handle all communication tasks. They need a set of cooperative protocols, sometimes called a protocol suite. Consider the following issues:

1. hardware failure—a host or router fails because either the hardware fails or the operating system crashes. The protocol software needs to detect such failures and recover from these failures if possible.

2. Network congestion – Even when all is working properly, networks have finite capacity. The protocol software needs to provide a way to suppress traffic in a congested machine.

3. Packet delay or loss—The protocol software needs to adapt to long delays.

4. Data corruption – The protocol software needs to detect and recover from such situations.

5. Data duplication or sequence errors – The protocol software needs to be able to reorder packets and remove duplicates.

If we look at how an analogous problem is answered by translation software, we see that program translation software breaks the problem into four parts: compiler, assembler, link editor, and loader. It should be clear that the pieces of the translation software must agree on the exact format of data passed between them. Thus, we see that the translation process involves multiple languages. This analogy holds for communication, as well, where the multiple protocols define the interfaces between the modules of communication. As the programming languages use a linear sequence in which the output of the compiler becomes input to the assembler, there is also a linear sequence for the communication software.

3.3 The Layers

Consider the modules of protocol software on each machine as being stacked vertically, like a stack of pancakes. Each layer takes responsibility for handling part of the problem.

Conceptually, sending a message from an application program on one machine to an application program on another means transferring the message the through successive layers of protocol software on the sender's machine, then transferring the message across the network, and finally transferring the message up through successive layers of protocol software on the receiver's machine.

In the real world, this is much more complex than this simple model makes things appear. Each layer makes decisions about the correctness of the message and chooses an appropriate action based on the message type or destination address. For example, one layer of the receiving machine must decide whether to keep or forward the message to another machine. Another layer must decide which application program should receive the message.

Let us consider the following model, which looks at the layers of protocol software used by a message that spans three networks; we understand that only the network interface and Internet Protocol layers need be considered since only those layers are needed to receive, route and then send datagrams. We understand that any machine attached to two networks must have two network interface modules, even though the conceptual thought shows only a single network interface layer in each machine. So, a sender on the original machine transmits a message that the IP layer places in a datagram and sends across the first network. On intermediate machines, the datagram passes up to the IP layer which routes the message back out again (on a different network). Only when the message reaches the final destination machine, does IP extract the message and pass it up to the higher layers of protocol.

3.4   *The disadvantage of layering*

Unfortunately, layering can be extremely inefficient, because of the division of the problem into discrete sub-problems. As an example, consider the job of the transport layer. It must accept a stream of bytes from an application program, divide the stream into packets, and send each packet over the internet. To optimize transfer, the transport layer should choose the largest possible packet size to allow one packet to travel in one network frame. If the software preserves strict layering, the transport layer cannot know how the internet layer will understand the datagram or frame formats--- thus, preventing an optimal transfers. Usually, practice will allow relaxing the strict layering to get around these and similar problems.

3.5   The need

In the discussion of client-server, and in particular, of message passing, there was mention of the action of sending messages from client to server (or vice versa). What was said was that it occurs as a result of a number of actions; what was glossed over earlier was how each of these individual acts gets done. People like to use analogies to explain a number of events in terms that they feel comfortable using. And, it isn't surprising that voice communications between people is a tempting model. The problem is that the voice model doesn't give an accurate analogy for electronic communications addressed here. We must really go back to scratch in order to derive the computer network issues.

Typically, there are a number of paths to and from a particular computer. But, this isn't all there is to communications in a computer network. What follows is an idea of what tasks need consideration.

1. the source system (or, transmitter) must not only enable the specific communications path for the transaction, but also make explicit mention to the network being transited of the destination station to which communication is being sent

2. the transmitting system must have a process by which it can find out if the receiving station not only is active, but also capable of accepting information

3. the "application" of the transmitting system or station must be able to knowing that the correct "application|" of the receiving station can be enabled and perform its functionality

4. incompatibilities between the transmitter and receiver must be capable of being resolved.

These tasks require a process to be enacted both in transmitter and receiver, not unlike the client-server process. Further, what happened in client-server was having functions having some amount of synergy between them were grouped into sets of similar functionality. This idea of constructing groups or sets of function to do a piece of functionality of the communications is precisely the concept of the protocol architecture.

From this very general context, we could construe the general sets of functions for communication to be:

1. process of transiting the network

2. process of computer access to the network

3. process of assuring integrity of communication

4. process of logical connectivity to the application.

The process of transiting the network, or "routing", is comprised of those functions which deal with the addressing of the message at the transmitter, and how the message is acted upon at each node in the network, as the message goes from transmitter to receiver. For example, the functions of constructing the address for the HDLC frame, finding what the address is and what routing mechanism will be used in routing would potentially reside in this process.

The process of computer access to the network may remind one of some of the work in the earlier chapters. The function of converting analog to digital signals or the digital-to-digital translations, the function of aligning the synchronization between transmitter and receiver, and so on, are potential functions comprising this process. The process is concerned with the exchange of data between computer and network,

The flow control and error detection/correction typify functions that could reside in the "integrity" layer. This layer concerns itself with reliable transfer of data. This layer can be implemented in many ways and is known as the "transport" layer.

Finally, the process of logical connections between applications deals with the applications themselves. One useful invention used here is the "Service Access Point" (SAP) . This concept arises from the way the client sends a request to the server, where the message has precise knowledge as to which application and the location of this application within the server. The message, from client-server, needs two, rather than one, address: 1) the physical (unique) address that allows the client to specify the specific server, and 2) the address of the specific application in the server's memory. This is directly analogous to calling each application by a logical name, so as to keep it separate from all other applications on the server.

As mentioned, the allocation of functions to a particular set has some tie-in with computer science. At the time when protocols were first being discussed, the software methodologies were also being discussed – specifically, modular programming. The idea of modules in software code is applicable here. This has been borne out in protocol implementations. For example, TCP/IP has perhaps 300 implementations worldwide, all of which will work together. These different implementations range between 6000 and 40,000 lines of equivalent C code.

Two standard protocols exist today. They are : 1)Transmission Control Protocol/Internet Protocol(TCP/IP), and 2) Open Systems Interconnections(OSI).

Work continues on them, albeit the OSI is really a conceptual model (although there are a few worldwide implementations.

3.6   TCP/IP

TCP/IP represents the protocol that enables communication on both the Internet and the World Wide Web. Historically, Vincent Cerf and Bob Kahn were the key creators of TCP/IP and the two first implementers of TCP/IP products. Its specifications are not standards, however, the specifications are kept and modified by the IETF.

From an overall perspective, TCP/IP has been loosely organized into groupings of functions, or layers, as follows

1. transport layer

2. internet layer

3. network access layer

4. application layer

5. Physical layer

The application layer and physical layer provide functionality, similar to all protocols. The "physical layer" deals with functions that pertain to the characteristics of the transmission medium, signals across the medium, and so on. The "application layer" is comprised of those functions with which the user has interest. These two layers have no further interest.

What differentiates TCP/IP from other protocol architectures is the remaining three layers.

Within the "Transport" layer, the objective is to make the data received as identical as possible with the data transmitted. We note this integrity-reliability issue is independent of the application. Typically, this grouping uses the TCP portion of the protocol. The "internet" layer deals with procedures of connecting devices in different networks. Two issues arise: 1) the process of routing data across different networks, and 2) the process of exchanging data at the point of interconnection of the different networks. This second issue infers that if the networks at the point of interchange have different protocols, a translation is required in order for the data to transit over the second network. The function solving this is that of a 'router'

The third functional grouping is the "network access" layer. The functionality deals with the interchange of data at the point between the computer and the network to which it is attached. For example, the type of service requested by the end user implies that different software be invoked in the network to enable a specific service (e.g., circuit switching or packet switching). Also of interest in this layer are the access functions- those doing analog- to-digital conversion, for example. We note that such functions are relatively independent of the specific network to which the computer is attached.

3.7   OSI

The OSI (Open Systems Interconnection) model, a standard in the international computer and telecommunications standards bodies, is much like TCP/IP in purpose; its difference arises from the very different implementation of the grouping of functions, and, in some cases, the specific functions within the layers.

OSI has seven layers, organized as follows:

1. Application
2. Presentation
3. Session
4. Transport
5. Network
6. Data Link
7. Physical

As with the TCP/IP model, the "Application" and "Physical" layers identify groupings of functions beyond the scope of this book, as they deal with the user application and medium issues, respectively.

The Presentation layer deals with providing an independent methodology for different applications in the sense that representations of data can have differences (i.e., EBSIDIC and ASCII have different representations). The Sessions layer provides structure for communications between applications without using the traditional communications structure of this book. Both of these layers are beyond the scope of this book.

The three layers dealing with communications across a network are 1) Data Link, 2) Network, and 3) Transport.

1. The Data Link layer provides for the reliable transfer of information across the Physical link. This layer also sends frames of user information with additional data, such as synchronization, error control and flow control across the physical link. This layer resembles the Transport layer of the TCP/IP protocol, in that some of the reliability issues of a point-to-point nature are resolved by OSI in the Data Link layer.

2. The Network layer of OSI provides upper layers with independence from the data transmission and switching technologies involving connected networks. The Network layer of OSI is responsible for establishing, maintaining and terminating circuits. The routing function (the IP layer of the TCP/IP protocol) is in the Network layer of OSI.

3. The Transport layer provides reliable and transparent transfer of data between end points, which can be different subnetworks. End-to-end error recovery and flow control are provided in Transport. Error recovery and flow control have been enhanced by the OSI design. For example, explicit routing and virtual route control are provided here.

As mentioned, the Session layer provides the control structure for communicating between applications. In this sense, establishment, termination and on-going management of active connections between cooperative applications is accomplished. This particular feature is used in the provision of ISDN (narrowband) service. Whereas this feature is implemented widely across Europe, penetration in the U.S. has been limited.

Chapter 4
TCP/IP and OSI

4.1   Introduction

We now know that the physical network addresses are both low-level and hardware dependent and we understand that each machine using TCP/IP is assigned one or more 32-bit IP addresses that are independent of the machine's hardware addresses. The same principle applies to 128-bit addresses. Application programs always use the IP address when specifying a destination. Hosts and routers must use physical addresses to transmit datagrams across underlying networks; they rely on address resolution schemes to perform the binding.

Usually, the machine's IP address is kept on secondary storage where the operating system finds it at startup. A question arises, " how does a machine without a permanently attached disk determine its IP address?" The problem is critical for workstations that store files on a remote server because such machines need an IP address before they can use standard TCP/IP file transfer protocols to obtain their initial boot image.

Because an operating system image that has a specific IP address

19

bound into the code that cannot be used on multiple machines, designers normally try to avoid compiling a machine's IP address in the operating system code. In particular, the bootstrap code is usually built so the same image can run on many machines. When such code starts execution, it uses the network to contact a server to obtain the machine's IP address.

The idea behind an IP address is simple: a machine that needs to know its internet address sends a response on a server on another machine, and waits until the server sends a response. We assume the server has access to a disk where it keeps a database of internet addresses. In this request, the machine that needs to know it internet address must uniquely identify itself, so the server can lookup the correct internet address and send a reply. Both the machine that issues the request and the server that responds use the physical network addresses during their brief communication.

Whenever a machine broadcasts a request for an address, it must uniquely identify itself. Any unique hardware identification works in this case.

4.2   *Reverse Address Resolution Protocol(RARP)*

The designers of TCP/IP protocols realized that there is another piece of uniquely identifying information readily available, namely the machine's physical network address. Using the physical address as an unique identification, has two advantages: 1) such addresses are always available, and 2) all machines on a given network will supply uniform and unique addressing.

A diskless machine uses a TCP/IP internet protocol called RARP to obtain its IP address from a server. In practice, the RARP message sent to request an internet address in that it allows a machine to request the IP address of a third party as easily as its own.

Let's see how a host uses RARP. The sender broadcasts a RARP request that specifies itself as both the sender and target machine, and supplies its physical network address in the target address field. All machines on the network receive the request, but only those authorized to supply the RARP service process the request and send a reply. For RARP to succeed, the network must contain at least one RARP server. Servers answer requests by filling in the target protocol address field, changing the message type from *request* to *reply*, and sending the reply back directly to the machine making the request. The original machine receives replies from all RARP servers, even though the first is needed.

4.3   A virtual Network

A user thinks of an internet as a single virtual network that interconnects all hosts, and through which communication is possible; its underlying architecture is both hidden and irrelevant.

4.4   TCP/IP Services

TCP/IP provides two services: 1) a connectionless packet delivery service at the lowest level, 2) a reliable transport service on which applications depend.

4.5   Concept of Unreliable Delivery

Internet software is designed around three conceptual networking services arranged in a hierarchy; much of its success has resulted because this architecture is surprisingly robust and adaptable.

The most fundamental internet service consists of a packet delivery system. In other words, the service is defined as an unreliable, best-effort connectionless packet delivery system, analogous to the service provided by network hardware that operates on a best-effort delivery paradigm. The service is called *unreliable* because delivery is not guaranteed The packet may be lost, duplicated, delayed, or delivered out of order. The service is called "connectionless" because each packet is treated independently from all others. A sequence of packets sent from one computer to another may travel over different paths, or some may be lost while others are delivered. Finally, the service is said to use "best-effort" delivery because the internet software makes an earnest attempt to deliver packets.

4.6 The Internet Datagram

On a physical network, the unit of transfer is a frame that contains a header and data, the header gives information such as the (physical) source and destination addresses. The internet calls its basic transfer unit an "internet datagram" also known as a "IP datagram".

The Internet datagram consists of the following fields:

1. VERS ; 4-bit field ; version of the IP protocol

2. HLEN ; 4-bit field ; gives the datagram header length

3. TOTALLENGTH ; length of IP datagram in octets

4. SERVICETYPE ; 8-bits ; 5-subfields specifying the transport

There are other fields, but we should mention the encapsulation in that more limits on frame size arise in practice. To make internet transport efficient, we guarantee that each datagram travels in a distinct physical frame.

Should a datagram be reassembled after passing across one network, or should the fragments be carried to the final host before reassembly? In a TCP/IP internet, once a datagram has been fragmented, the fragments travel as separate datagrams all the way to the ultimate destination where they must be reassembled.

Chapter 5

Routing in a TCP/IP environment

5.1 Routing in an Internet

In a packet switching system, "routing" refers to the process of choosing a path over which to send packets, and "router" refers to a computer making such a choice. Routing occurs at several levels. For example, within a wide area network that has multiple physical connections between packet switches, the network itself is responsible for routing packets from time to time they enter until they leave. Such internal routing is completely self-contained inside the wide area network. Remember that the goal of IP is to provide a virtual network that encompasses multiple physical networks and offers a connectionless datagram delivery service. So, we focus on IP routing.

Routing in an internet can be difficult, especially among computers that have multiple physical network connections. Ideally, the routing software would examine such things as network load, datagram length, or the type of service specified in the datagram header, when selecting the best path. Practically, this doesn't occur.

To understand IP routing, we go back and look at the architecture of a TCP/IP internet. Recall that an internet is comprised of multiple physical networks interconnected by computers called "routers". Each router has direct connections to two or more networks. By contrast, a host computer usually connects directly to one physical network.

Both hosts and routers participate in routing an IP datagram to its destination. When an application program on a host attempts to communicate, the TCP/IP protocols eventually generate one or more IP datagrams. The host must make a routing decision when it chooses where to send the datagrams. The hosts must make routing decisions even if they only have one network connection.

5.2   Direct and indirect Delivery

We know that one machine on a given physical network can send a physical frame directly to another machine on the same network. To transfer an IP datagram, the sender encapsulates the datagram in a physical frame, maps the destination IP address into a physical address, and uses the network hardware to deliver it. So, in summary, transmission of an IP datagram between two machines on a single physical network does not involve routers. The sender encapsulates the datagram in a physical frame, binds the destination IP address to a physical hardware address and sends the resulting frame directly to its destination. Since we know the IP addresses are divided into a network-specific prefix and a host-specific suffix, the sender extracts the network portion of the destination IP address and compares it to the network portion of its own IP address(es). A match means the datagram can be sent directly. One of the Internet advantages is that because the network addresses of all machines on a single network include a common network prefix, and because extracting that prefix can be done in a few machine instructions, testing whether a machine can be reached directly is extremely efficient.

Indirect delivery is more difficult than direct delivery because the sender must identify a router to which the datagram can be sent. Essentially, routers in a TCP/IP internet form a cooperative, interconnected structure. Datagrams pass from router to router until they reach a router that can deliver the datagrams directly.

5.3   Table driven IP routing

The usual IP routing algorithm employs an Internet Routing Table on each machine that stores information about possible destinations and how to reach them. Because both hosts and routers route datagrams, both have routing tables. Whenever the IP software in a host or router needs to transmit a datagram, it consults the routing table to decide where to send the datagram. Conceptually, we would like to use the principle of information hiding and allow machines to make routing decisions with minimal information. Fortunately, the IP address scheme helps achieve this goal. Recall that IP addresses are assigned to make all machines connected to a given physical network share a common prefix (the network portion of the address) . We also recall that such an assignment makes the test for direct very efficient. It also means that routing tables only need to contain network prefixes and not full IP addresses.

5.4   Next-hop routing

Using the network portion of a destination, instead of the complete host address, makes routing efficient and keeps routing tables small. More important, it helps hide information, keeping the details of specific hosts confined the local environment in which these hosts operate. Typically, a routing table contains pairs (N,R), where N is the IP address of a destination network, and R is the IP address of the "next" router along the path to network N. Router R is called the 'next hop, and the idea of using a routing table to store a next hop for each destination is called "next-hop routing" . Thus, the routing table in a router R only specifies one step along the path from R to a destination network-the router does not know the complete path to a destination. The underlying principle can be summarized as follows:

To hide information, keep routing tables small, and make routing decisions efficient, IP routing software only keeps information about destination network addresses, not about individual host addresses.

Another way to hide information and keep routing tables small consolidates multiple entries into a default case. The idea is to have the IP routing software first look in the routing table for the destination network. If no route appears in the table, the routing tables send the datagram to a 'default router'.

5.5   The IP routing algorithm

Algorithm: RouteDatagram (Datagram, RoutingTable)

Extract destination IP address, D from the datagram and compute the network prefix, N;

If N matches any directly connected network address, deliver datagram to destination D over that network

(This involves resolving D to a physical address, encapsulating the datagram, and sending the frame.)

Else if the table contains a host-specific route for D send datagram to next-hop specified in table;

Else if the table contains a route for network N send datagram to next-hop specified in table;

Else if the table contains a default route send datagram to the default router specified in table;

Else declare a routing error.

5.6   Routing with IP addresses

It is important to recognize that except for decrementing the time-to-live parameter and recomputing the checksum, IP routing does not alter the original datagram. In particular, the datagram source and destination addresses remain unaltered; they always specify the

IP address of the original source and the IP address of the ultimate destination. When IP executes the routing algorithm, it selects a new IP address, the IP address of the machine to which the datagram should be sent next. The new address is most likely the address of a router.

We said that the IP address selected by the IP routing algorithm is known as the 'next hop' address because it tells where the datagram must be sent next. After executing the routing algorithm, IP passes the datagram and the next hop address to the network interface software responsible for the physical network over which the datagram must be sent. The network interface software binds the next hop address to a physical address, forms a frame, and sends the result. After using the next hop address to find a physical address, the network interface software discards the next hop address.

IP dutifully extracts the destination address in each datagram and uses the routing table to produce a next hop address. It then passes the datagram and next hop address to the network interface, which recomputes the binding to a physical address. If the routing table used physical addresses, the binding between the next hop's IP address and physical address could be performed once, saving unneeded computation.

The IP software avoids using physical addresses when storing and computing routes for two reasons; 1) the routing table provides an especially clean interface between IP software that routes datagrams and high-level software that manipulates routes, and 2) the whole point of the Internet Protocol is to build an abstract that hides the details of the underlying networks.

5.7   Handling Incoming Datagrams

We have discussed outgoing packets. It should be clear that the IP software must process incoming packets as well.

When an IP datagram arrives at a host, the network interface software delivers it to the IP software for processing. If the datagram's destination address matches the host's IP address, IP software on the host accepts the datagram and passes it to the appropriate higher-level protocol software for further processing. If the destination IP address does not match, a host is required to discard the datagram.

Unlike hosts, routers perform forwarding. When an IP datagram arrives at the router, it is delivered to the IP software. Again, two cases arise: 1) the datagram could have reached its final destination, or 2) it may need to go further. If the datagram destination IP address matches the router's own IP address, the IP software passes the datagram to higher-level protocol software for processing. If the datagram has not reached its final destination, IP routes the datagram using the standard algorithm and the information in the local routing table.

There are four reasons why a host not designated to serve as a router should refrain from performing any router functions: 1) when such a host receives a datagram intended for another machine, an error has occurred with internet addressing, routing, or delivery; 2) routing will cause unnecessary network traffic ; 3) simple errors cause chaos; and 4) routers do more than merely route traffic.

Chapter 6

Internet Security and Firewall Design

6.1   Introduction

Like the locks used to help keep tangible property secure, computers and data networks need provisions that help keep information secure. Security in an internet enviroment is both important and secure. Security in an internet is difficult because security involves understanding when and how participating users, computers, services, and networks can trust one another, as well as understanding the technical details of network hardware and protocols. More important, because TCP/IP supports a wide diversity of users, services, and networks, and because an internet can span many political and organizational boundaries, participating individuals and organizations may not agree on a level of trust or policies for handling data.

6.2   Protecting Resources

The terms, "network security" and "information security" refer broadly to confidence that information and services available on a network cannot be accessed by unauthorized users. Providing security for information requires protecting both physical and abstract resources. Physical resources include passive storage devices such as user's disks. In a network environment, physical security extends to cables, bridges and routers. Protecting abstract resources such as information is usually more difficult as information is more elusive. Data Integrity (i.e., protecting information from unauthorized change) is crucial; so is data availability (i.e., guaranteeing outsiders cannot prevent legitimate data access by flooding a network with traffic). Security needs to be usually more restrictive.

6.3   Mechanisms for Internet Security

Authentication Mechanisms

"Authentication Mechanisms" solve the problem of verifying identification. Many servers, for example, are configured to reject a request unless the request originates from an authorized client. When a client first makes contact, the server must verify that the client is authorized before granting service. To validate authorization, a server must know the identity of a client. When using IP address authentication, a manager configures a server with a list of valid IP source addresses. The server examines the source IP address on each incoming request and only accepts requests from client computers on the authorized list. Note that an authentication scheme that uses a remote machine's IP address to authenticate its identity doesn't prevent attacks by imposters across an unsecured internet because an imposter who gains control of an intermediate router can impersonate an authorized client.

We can avoid this problem by using a trusted service. One form of trusted service uses a 'public key encryption' system. To use such a system, each participant must be assigned two 'keys' that are used to code and decode messages. A participant publishes one key, called a 'public key', in a public database and keeps the other key secret. A message encoded using one key, can be decoded by the other. A client and server that use public key encryption can be reasonably sure that the other communicant is authentic, even if datagrams transferred between them pass through an unsecure network.

6.4   Firewalls and Internet access

Mechanisms that control internet access handle the problem of screening a particular network or an organization from unwanted communication. Unlike authentication and privacy mechanisms, which can be added to application programs, internet access control usually requires changes to the basic components of the internet infrastructure. In particular, successful access control requires a careful combination of restrictions on network topology, intermediate information staging and packet filters.

A single technique has emerged as the basis for internet access control. The technique places a block known as an internet 'firewall' at the entrance to the part of the internet to be protected. A firewall partitions an internet into two regions, referred to as the 'inside' and 'outside'. An organization that has multiple external connections and must coordinate all firewalls; failure to restrict access identically on all firewalls can leave the organization vulnerable.

To be effective, a firewall that uses datagram filtering should restrict access to all IP sources, IP destinations, protocols, and protocol ports except those computers, networks, and services the organization explicitly decides to make available externally. A packet filter that allows a manager to specify which datagrams to admit instead of which datagrams to block can make such restrictions easy to specify.

Chapter 7

Error and control messages

7.1　Introduction

If a router cannot route or deliver a datagram directly a datagram, or if the router detects an unusual condition that affects its ability to forward the datagram, the router needs to inform the original source to take action to avoid or correct the problem. We will see that routers use the mechanism to report problems and hosts use it to test whether destinations are reachable.

The mechanism, known as the 'Internet Control Message Protocol', (ICMP), allows routers to send error or control messages to other routers or hosts; ICMP provides communication between the Internet Protocol software on one machine and the Internet Protocol software on the other.

7.2　Error Reporting vs. Error Correction

Technically, ICMP is an error reporting scheme. It provides a way for routers that encounter an error to report the error to the original source. When a datagram causes an error, ICMP can only report the error condition back the original source of the datagram; the source must relate the error to an individual application program or take other action to correct the problem.

7.3   ICMP Message Delivery

Each ICMP message travels across the internet in the data portion of an IP datagram, which itself travels across each physical network in the data portion of a frame. Datagrams carrying ICMP messages are routed exactly like datagrams carrying information for users; there is no additional reliability or priority. It is important to recall that even though ICMP messages are encapsulated and sent using IP, ICMP is not considered a higher-level protocol—it is a required part of IP.

Although each ICMP message has its own format, they all begin with the same three fields: an 8-bit integer message TYPE field that identifies the message, an 8-bit CODE field that provides further information about the message type, and a 16-bit ICMP CHECKSUM field.

Chapter 8

Routers and Interworking

8.1 Introduction

In the scheme of things, a capability is needed to transmit across different networks. Typically, we may define something to be a network, such as the Internet, which is actually comprised of many individual networks, each connected to each other. The process of using a protocol architecture to connect nodes of different networks for the purpose of exchanging information is known as 'interworking'. As one might suspect, there are devices to perform this work (i.e., hardware and software platforms. This device, providing a communications path and needed logic to exchange information between nodes in different sub-networks, is called a router.

A bridge is a device involving communications across a single type of network. In order to achieve communications across different networks, the functions that the router requires include the following:

1. provide a link or route between sub-networks

2. provide for routing and delivery of data between

applications residing on end systems, which themselves are connected to different sub-networks

3. provide the functionality independently of the architecture of the attached sub-networks.

4. Addressing scheme

5. Maximum packet sizes

Chapter 9

The Web and its relation to client-server and Protocol Architecture

9.1 Introduction

The World Wide Web (or, the "Web") is an invention that has experienced unimaginable growth ever since the late 1990s; it is the engine that drives much e-commerce and computer science of today's world. Most observers believe that this growth will continue for the foreseeable future, and the invention will impact us as few, if any, inventions have over the history of humankind. Whether or not the "Web" in its current implementation succeeds, there will be a network structure, providing the kind of functionality as the web currently provides.

This chapter considers the evolving network structure of the Web, and postulates that the basic infrastructure of this structure is already in hand, from the principles and protocols, discussed earlier in this book.

The basic idea of the Web is that pages of information are shared; these pages are "linked", or joined, in some fashion. The end user has access to the web and these pages through a 'browser', which is a software program, running on the end user' s computer. This browser connects the end user to the Internet and sends information to the desired Web server, containing these pages. The Web server accesses these pages, which are either locally or remotely stored, and requests that these pages be made available to the end user. This results in the requested pages being sent to the end user. Tim Berners-Lee has defined the browser as a web client. We begin to understand the end-user perspective as a client-server application where information happens th be in the from of pages (i.e., specially encoded pages). We now look at the detailed components.

Underlying the physical components, there are some basics about the Web. The pages of the Web are composed in the hypertext markup language (i.e., HTML). Each page is identified by an address; this address is the URL (i.e., Uniform Resource Locater). This address is unique across all pages linked on all servers across the Internet. That these issues were made standard is perhaps the crucial technical achievement of the Web. Using client-server language to discuss how the Web components relate, we see that the browser software resembles operating system functions of Remote Call Procedure. Such functions form a message, which is to be transmitted to the server. Communication between browser and server occurs in transactions in the form of frames. For the Web, the communications protocol is the TCP/IP protocol; specifically the routing function of TCP is employed in the transfer of the browser request to the server. The end user is interested in getting "pages", or information, which is displayed through the browser through the HTML language. The HTTP (or Hypertext Transfer Protocol) is used to perform the host-to-host integrity function, so that the packets of data, routed by TCP, can be integrated into the page structure. In addition, HTTP instructs the browser to display the page on the display device, using the HTML language. TCP/IP essentially sends packets which are formed into pages by the HTTP protocol and the page is translated by HTML into the page on the display. Thus, the client-server model is the underlying infrastructure of the Web, the protocol issues of separating functionality into different layers for Web communication, is directly analogous to the earlier chapters.

Again, the standardization of the language that pages would be written in HTML, the unique address so that individual pages can be found (URL), and the protocol glue to take packets that have been communicated through TCP/IP and form pages (through HTTP) is the infrastructure of the Web. This doesn't mean that something else couldn't be substituted for the requisite functions. And, it shouldn't be surprising that groups have been and continue to work on enhancements of the Web for future implementations.

Epilog

SO, WE COME TO the conclusion. We have seen how the principles, the protocols, and the features are all equally important to the story. The ideas of hiding details, of partitioning functions in protocols, and of layering of functionality in the protocols, are important to networks. I have seen a parallel of these principles to the formation and rise of the idea of interchangeable parts that underlies the Industrial Revolution. In a sense, the hiding details, the partitioning of functions, and the layering of functionality become the glue between our computer portion, or the interchangeable paths analogy to the Industrial Revolution.

My personal journey through all this is due to a number of individuals, some commented on before, and others who it has been a privilege to know. Thanks to them all.

References

ABRA85 Abramson,N., "Development of the ALOHANET", IEEE Transactions on Information Theory, March 1985.

ASH90 Ash,G.,"Design and Control of Networks witn Dynamic Nonhierarchical Routing", IEEE Communications Magzine, October 1990.

BELL90 BELLCORE, Telecommunications Transmission Engineering, three volumes, 1990

BERN96 Bernstein, P., "Middleware: A Model for distributed System Services", Communication of the ACM, February 1996.

BEYD00 Beyda, W., Data Communications: from Basics to Broadband, Prentice-Hall,2000.

BLAC93 Black,U., Data Link Protocols, Prentice-Hall, 1993.

BLAC96 Black, U., Physical level Interfaces and Protocols, IEEE Computer Society Press, 1996.

CASA94 Casavant, T., and Singhal, M.,eds. Distributed Computing Systems, IEEE Computer Society Press, 1994.

CERF94 Cerf, V., and Kahn,R., " A Protocol for Packet Network

Interconnection", IEEE Transactions on Communications, May 1974.

COME99 Comer,D., "Computer Networks and Internets", Prentice-Hall, third edition .

COME95 Comer, D. ," Internetworking with TCP/IP", Prentice-Hall, third edition, 1995.

DEWI93 Dewire, D., "Client/Server Computing, Mcgraw-Hill, 1993.

DODD00 Dodd, A., "The Essential Guide to Telecommunications", 2nd Edition , Prentice-Hall, 2000.

DRUC89 Drucker,P., "The New Realities", Harper & Row, 1989.

EVAN00 Evans, P., and Wurster,T., "Blown to Bits", Harvard Business School Press, 2000.

FREE94 Freeman, R., Telecommunications Transmission Handbook, Wiley, 1994.

HARB92 Harbison. R., "Frame Relay: Technology for our Time", LAN Technology, December 1992.

HELD01 Held,G., "Data over wireless Networks", McGraw-Hill, 2001.

HELG91 Helgert, H.,Integrated Services Digital Networks: Architectures, Protocols,and Standards. Addison-Wesley, 1991.

HOUS01 Housel, T., and Skopec, E.W., Global Telecommunications Revolution-the Business Perspective, McGraw-Hill, 2001.

JOHN01 Johnson, P., Introduction to Networks and Telecommunications , iUniverse Press, 2001,

JOHN94 Johnson,P., "Domestic and International Standards" , Telecommunications Network Management into the 21st Century, Aidarous and Plevyak, Eds. , IEEE Press, 1994.

JOHN04 Johnson, P., "Introduction to Business Data Communications with wireless and Broadband", iUniverse Press,2004.

KUMA95 Kumar, B.," Broadband Communications: A Professional's Guide to ATM, Frame relay, etc., McGraw-Hill, 1995.

MADR94 Madron, T., "Local Area Networks:New Technologies, Emerging Standards; Wiley, 1994.

MART90 Martin, J., Telecommunications and the Computer, Prentice-Hall, 1990.

MART94 Martin, J., and others," Local Area Networks ", Prentice-Hall, 1994.

MOUL01 Moulton, P., "The Telecommunications Survival Guide", Prentice-Hall, 2001.

PANK99 Panko, R., Business Data Communications and Networking, Prentice-Hall, 1999.

PORT80 Porter, M., Competitive Strategies, The Free Press, 1980.

PORT85 Porter, M., Competitive Advantage, The Free Press, 1985.

RAPP96 Rappaport, T., Wireless Communications, Prentice-Hall, 1996.

RENA96 Renaud, P., An Introduction to Client/Server Systems, Wiley, 1996.

STAL95 Stallings, W., ISDN and Broadband ISDN, with Frame Relay and ATM, Prentice-Hall, 1995.

STAL97 Stallings, W., Local and Metropolitan Networks, Prentice-Hall, 1997.

STAL01 Stallings , W., Business Data Communications, fourth edition, Prentice-Hall, 2001.

TANE01 Tanenbaum, A., Computer Networks, fourth edition, Prentice-Hall, 2001

WHET96 Whetzel, J., "Integrating the World Wide Web and Database Technology", AT&T Technical Journal, March/April 1996.

www.ingramcontent.com/pod-product-compliance
Lightning Source LLC
Chambersburg PA
CBHW051215050326
40689CB00008B/1323